Everything There Is

in the whole of the universe

(The Theory of Unified Absolute Relativity)

Introduction

The trouble with Grand Unifying Theories is - they're just too damned grand; I mean too cumbersomely cobbled together. Time and again I have read that String Theory (see Appendix Two) is most likely to provide unification of all the forces in physics. But String Theory and its ugly sisters wobble on and on, growing ever more complex and convoluted as they are modified and adapted to fit their own fantasies. My purpose in writing this little book is to propose a much less grand, yet absolute unification.

For years, I kept thinking that there had to be a better, less fragmented description of the fundamental particles and forces in nature. And by that, I mean a more simple solution, able to answer all of the big questions left unsolved by the mysteries inherent in Quantum Mechanics as it stands today.

As particle physics has developed, along with Quantum Mechanics, it has met itself coming the other way so often, "discovering" more and more new particles, theorising on even more, constructing parallel universes, quantum states of superposition with everywhere-existing particles, while failing to adequately describe the quantum's relationship with the world in which we live. All this, without a coherent description of the electron.

Nuclear physics set out to describe what is at the "heart" of everything – how basic particles react with one another, creating everything we see, everything in our world, the stars and the universe. It has failed. Many physicists have thrown in the towel, insisting now that such a comprehensive description is and always will be impossible. String theorists have picked up that towel and tied it into impossibly tiny, unobservable knots.

Most physicists seem to agree that it is not possible to describe fundamental particles and forces in a "classical" way – that is, way that can be related directly to the world around us, using just our three spatial dimensions and time. Well I will show that it *is* possible – in fact, this little book will constitute just such a comprehensive description. It is all in here. This book holds the key to the fundamental laws governing the whole universe, and the one-and-only, one single particle upon which everything, absolutely everything is built.

This book then, with its Theory of Unified Absolute Relativity, its theory of everything, will not be so grand as to be incomprehensible to the interested layman; in fact, anyone and everyone should be able to understand the particles, the forces and interactions at the very heart of our wondrous universe – because, contrary to current scientific thinking, they are surprisingly simple.

Prologue

There are significant unanswered questions in physics – questions that many physicists are beginning to ignore or even to deny. For example, the wave/particle duality of matter. That is, looked at in one way, matter behaves as if it consists of particles, then, looked at in another, it appears to behave like a wave. Many physicists have stopped seeing this as an enigma, accepting that the duality is just the way it is. But then some explanations like to suggest that the particles form waves, so there is no question to be answered. But there is still no description as to how these particles arrange themselves in waves. In fact, when systems are constructed which produce a single particle at a time, over time the single particles still seem to perform as if they have been guided by some force to behave like waves. No one can explain this. Scientific investigation, or the failure of it, seems to be suggesting that such shortcomings in our understanding are simply to be accepted.

There are many other questions, so apparently unanswerable that many physicists believe that is will always be impossible to describe what is happening on a quantum level in "classical" terms; that is, in a non-Quantum-Mechanical, real-world way. Well I do not think this should be so. Neither is it necessary to give up looking for answers. And these answers should always, without exception, be in the form that anybody can understand, existing in three dimensions and time. So, if we answer all the questions, we will have acceptable explanations to all the mysteries – which also means describing gravity on a quantum level (something that has defied all previous attempts).

So, let us pose the questions first, or at least reference them in a simple and logical way – and then answer them. All of them.

There are a multitude of atomic and sub-atomic particles that have been identified (plus many more that have been speculated upon but never found – see Appendix Five), but my theory requires that we discuss, in the first instance, just the following six types of particles:

Protons, Neutrons, Electrons, Quarks, Photons and *Higgs bosons*:

Protons and *Neutrons* are the constituents of the nucleus of atoms. Neutrons do not have charge, which means they are electrically neutral (hence the name), but protons have charge.

Electrons "orbit" the nucleus of atoms (they don't actually orbit, but we will more thoroughly discuss this in the text body of this book). There are the same number of electrons orbiting a nucleus as there are protons in the nucleus.

Quarks are said to be the constituent parts of protons and neutrons. They are defined as fundamental particles, that is to say not made up of other particles. Quarks come in different types: "Up" and "Down" quarks, "Charm", "Strange", "Top" and "Bottom" quarks (please do not concern yourself with the titles of these things; their names are not significant). Protons are said to consist of two "up" quarks and one "down" quark, whereas neutrons are said to consist of one "up" quark and two "down" quarks.

Photons have been determined as particles of light and all the other types of electromagnetism (x-rays and so on).

Higgs bosons are said to be massive particles with no charge, responsible for giving other particles their mass.

The Method

To keep the process of the discussion as simple as possible, the method used in piecing together Unified Absolute Relativity is to pose the relevant questions and then proceed to answer them. Every one of them.

(There is one single unanswered question, which is given a place in the appendices, as you will see – but the question itself, as far as I know, has never been posed. I include it because it is a good question, to which I have no answer.)

I have included some mathematical equations in the text, but very few. Even then, they are there to illustrate that the mathematics behind the ideas exists; therefore, it is never necessary to understand the mathematics, as the ideas are discussed fully (of course, further study into all the ideas is then possible).

The Questions and The Solutions

All the explanations of the experiments and phenomena that highlight the questions in physics that have been included in this book are easily referenced against many other publications in print and on the internet. These experiments have been carried out on numerous occasions, with many papers, articles and books published on these matters. Anyone wishing to investigate these phenomena further than the outlines with which I have chosen to illustrate the questions posed, can find a wealth of information available, all of which will lead to the same fundamental, so-far unanswered questions.

To establish all the questions and their solutions, using Unified Absolute Relativity, we will need to discuss the following:

Section One – The Double-Slit Experiment (including The Copenhagen Interpretation and The Many-Worlds Theory)

Section Two – The Four Forces

Section Three – The Different Masses of Particles

Section Four – Quantum Entanglement

Section Five – Matter and Antimatter

The solutions can always be understood by anyone, with no specialist scientific or mathematical training, allowing a classical description of the quantum world and its logical interaction with the macro world in which we all live.

Appendices

This last section of the book discusses some of the topics from Questions and Solutions in more detail. We shall also examine other aspects of quantum mechanics that have not so for been touched upon in the first two parts, with special focus on the concept of Unified Absolute Relativity as herein previously determined and explained.

Contents

The Questions and Their Solutions

Appendices

Conclusion

Part One

The Double Slit Experiment - The Questions

This is the very famous and well-studied experiment that illustrates the wave/particle duality of protons and electrons. Appreciation of what this experiment shows is essential for the understanding of Unified Absolute Relativity.

The double-slit experiment, or sometimes called Young's interference experiment, is a simple set-up of light source shining through two slits onto a

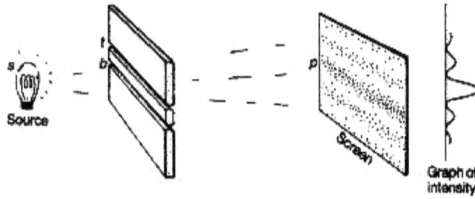

receptor screen as illustrated below.

What happens is that the two streams of light on the receptor side of the double-slit set-up cause a pattern on the screen of lighter and darker bands. This is because the light is coming in waves: so picture two wave sources in water, two separate, waggling fingers for example. The water waves will interfere with each other, with some parts forming troughs and cancelling each other out, some parts combining to form peaks, as illustrated below, which, you can imagine, would lead to an interference pattern on a screen, (see next page):

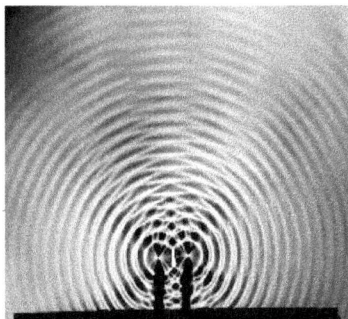

This shows the wave nature of light; in fact, coloured light is measured in its wavelength, or wave-frequency. Blue light is of shorter wavelength, changing more frequently, while red light is of longer wavelength, changing less frequently. The rapidity of change also denotes the *energy* of the light, whereas the intensity is the *amount* of light. Blue light is therefore more energetic than red light.

So, light is a wave.

But when the intensity of the light source is turned down low enough, just one "piece" of light goes through the system at a time, which instigates a single response at the screen. One piece of light, one quanta, is the photon, the single particle of light. Which suggests that light is actually a particle, not a wave. And the particles appear on the screen randomly; no matter how the responses are analysed, there is no predicting the next response position, no telling where the photon is likely to land on the screen.

The strange thing is, though, that if the system is left on like this, with single particles landing randomly on the screen, the pattern they make will eventually build back into the interference pattern, exactly as before.

So how do the random particles "know" where they should and should not go? Somehow, they are aware of where the other photons have landed, and are able to arrange themselves to finally build the interference pattern back again. How?

And then, perhaps even stranger, is what happens when physicists attempt to find out through which of the two slits the photon has passed. By placing just one detector into the system, at one slit or the other, so that the particle can be detected there or not, whenever a photon makes it to the screen physicists can tell if it has either passed through the slit with the detector or the other slit. The detector doesn't stop the particle – the photon still gets to the screen. But, guess what? The interference pattern has disappeared. The pattern appears as two separated light patches directly behind the two slits on the screen, as if light was not a wave at all.

Even if the detector is positioned in the system *after* the slits, the photons seem to "know" they are being examined and change their behaviour. In fact, it is possible to leave the detector in place but stop taking readings and the interference pattern returns. Examine the system and the pattern disappears again.

The system will simply not allow itself to be examined. The American physicist Richard P. Feynman called it *"a phenomenon which is impossible ... to explain in any classical way, and which has in it the heart of quantum mechanics. In reality, it contains the only mystery (of quantum mechanics)"*. And he was fond of saying that all of quantum mechanics can be gleaned from carefully thinking through the implications of this single experiment.

*   *   *

The implications of the double-slit experiment have been thought through many times and there have been many interpretations of what exactly is happening within this system. The most important and famous of which is:

The Copenhagen Interpretation

The Copenhagen Interpretation, first posed by physicist Danish physicist Niels Bohr in 1920, suggests that a quantum particle (photon or electron, perhaps) doesn't exist in one state or another, but in all its possible states at once. It's only when we observe its state that a quantum particle is forced to choose one probability, and that's the state that we observe.

This state of existing in all possible states at once is called an object's coherent superposition, or simply superposition of states. The total of all possible states in which an object can exist makes up the object's wave function (wave functions are discussed a bit later in this section). When we observe an object, the superposition collapses and the object is forced into only one of its states.

So, the *Copenhagen Interpretation* determines that the photon (or electron) exists everywhere until we decide to measure it. Therefore, it reasons, it must be the observer who is responsible for the actual positioning of the photon. In other words, the photon behaves as it does *because* we are examining it.

*   *   *

In response to the *Copenhagen Interpretation*, Viennese physicist Erwin Schrödinger in 1935, devised a famous thought experiment, ever since called *Schrödinger's Cat;* not to show how the quantum micro world relates to the macro world, but to illustrate the apparent impossibility of relating the standard model of quantum mechanics to the real world in which we live.

There are many versions if this experiment, but essentially it goes like this:

An unfortunate cat is locked in a box which has two connected compartments. The cat is trapped in one compartment and a gun is pointed at its head (just a thought experiment, remember). The gun is triggered by a device that will detect the presence of a single photon that is released into the system – just the one photon in the whole sealed box. Now, if the photon is in the compartment with the cat and the gun, the detector will sense it and the cat will die. If the photon is in the other compartment, the cat lives. But the photon will be in a superposition of states until we open the box and attempt to examine it.

So, is the cat dead or alive? If the photon is in a superposition of states, then the detector has detected the photon and it has not detected the photon, and the gun has and has not been fired. If the photon is in a superposition of states, then so is the cat. It is neither dead nor alive, or it is both, choose whichever one you fancy.

I'm sure you will agree, this thought experiment certainly does manage to show that the standard model of quantum mechanics disallows a simple and logical connection between the micro and the macro worlds.

*   *   *

Going a huge step further into implausibility, the *Many-Worlds Theory* suggests that as the photon exists everywhere until it is observed, it is fixed into a single position here only; but all the other infinite possibilities are still realised in an infinite number of parallel universes. In other words, every time a photon is positioned here, an infinite number of alternative photons are positioned individually in an infinite number of instantaneously-created alternative, parallel universes.

A significant number of scientists have taken up such far-flung suggestions as the *Many-Worlds Theory* amongst others – others which include *String Theory* and its sister *M Theory*. These theories depend upon other dimensions, sometimes many of them, even possible links through membrane-like parallel universes. But the questions asked in this way will always be far, far out of the reach of conclusive answers. All we should do, all we *can* do is to show why such suggestions and their attendant paradoxical questions are totally and irrefutably irrelevant.

*   *   *

Before we leave the double-slit experiment and concentrate on other questions, two other aspects should be taken into consideration - *Schrödinger's Wave Equation* and *Heisenberg's Uncertainty Principle*:

Schrödinger's Wave Equation

At the beginning of the twentieth century, experimental evidence suggested that atomic particles were also wave-like in nature. For example, electrons were found to give diffraction patterns when passed through a double slit in a similar way to light waves. Therefore, it was reasonable to assume that a wave equation could explain the behaviour of atomic particles.

Schrödinger was the first person to write down such a wave equation. Much discussion then centred on what the equation meant. After much debate, the wavefunction is now accepted to be a probability distribution with the wavefunction giving the probability of finding the particle at a certain position.

The time-dependent Schrödinger equation:

$$i\hbar\frac{\partial\psi(\mathbf{r},t)}{\partial t} = -\frac{\hbar^2}{2m}\nabla^2\psi(\mathbf{r},t) + V(\mathbf{r})\psi(\mathbf{r},t)$$

It doesn't matter if you do not understand the above equation – it has been included as an important piece of descriptive mathematics, as its intention is to determine the probability of finding an electron (or photon) at a point in space and time. The thing to notice is that we are talking only of probabilities, as nothing seems certain in the quantum world. Light photons and electrons fly around apparently everywhere all at once until a person looks at them. We can't tell how fast they're going or where they are, at least not both things at the same time. Which is what our other aspect is all about:

Heisenberg's Uncertainty Principle

The more precisely the position is determined, the less precisely the momentum is known in this instant, and vice versa. - Werner Heisenberg

A direct consequence of the dual (particle/wave) nature of all matter and energy is the uncertainty principle, which states that: 'You cannot know the position of a particle and how fast it's moving with arbitrary precision at the same moment.' Or, 'It is fundamentally impossible to simultaneously know the position and momentum of a particle at the same moment with arbitrary accuracy.'

Quantitatively, the principle can be stated as follows:

$\Delta x.\Delta p \geq h/2\pi$

(where Δx is the uncertainty in position, Δp is the uncertainty in momentum and h is a constant called Planck's constant)

So, take at least this sentence from the above: 'You cannot know the position of a particle and how fast it's moving with arbitrary precision at the same moment.'

*   *   *

All this means to say that photon particles are bits of light that are also waves of light, they seem to know when they're being looked at and alter their behaviour if they are, they only have a probability of being in a certain place at a certain time and you can't find out how fast a particle's going if you know where it is, and if you know how fast it's going you can't know where it is – because perhaps it's in other universes, infinitely many of them.

At this stage, it's worth repeating the following:

The double-slit experiment has become a classic thought experiment for its clarity in expressing the central puzzles of quantum mechanics. Because it demonstrates the fundamental limitation of the ability of the observer to predict experimental results, Richard Feynman called it "a phenomenon which is impossible ... to explain in any classical way, and which has in it the heart of quantum mechanics. In reality, it contains the only mystery (of quantum mechanics)" And he was fond of saying that all of quantum mechanics can be gleaned from carefully thinking through the implications of this single experiment.

Well, we will think it through – and then we shall explain it, all of it, in a classical (that means, in this case, a logical, conclusive and non-confusing) way.

The Double Slit Experiment - The Solutions

So, if we can offer a solution to the problems posed by the double-slit experiment, we will have made a significant breakthrough. One evening I held my hand under the table lamp – the light faded as I moved my hand away and grew more intense as I got closer to the light source. All very simple. But why did it do this? If photons of light were coming straight out of the lamp, how did the particles grow smaller to become less dense when covering my illuminated skin as I moved my hand away from the lamp?

Then I realised something – and this was the whole essence of whatever was to follow: I saw that nothing had happened until the light reached my hand. I mean that, in effect, it wasn't there. The light, and the heat, was an effect – but my hand was causing the effect.

When I took this thinking to the double-slit experiment, I was astonished to see the answer to the paradox of wave/particle duality and to all the other questions posed by this experiment. The light left the light source as a wavefunction and it "spread out" as it went, according to the inverse square of the distance – but nothing happened to it until it got to the screen. The double slits fashioned the light into a shape that made the waves "interfere" with one another, and the screen responded to that pattern – that is, the particles of the screen responded, depending upon which fixed screen particle was susceptible (see Appendix One) at which time. Effectively, there were no photons, no pieces of light, but a pattern of waves. The pieces of light were the matter particles of the screen responding. Light is a wave operating on a receptor made of particles, the screen, the atoms of which jump from one energy level to another, giving the *impression* of the single quanta or particle of light.

What happens when we dim the light is that the interference pattern remains, where each wave interferes with itself, and only with itself; but now it has the power to influence on the receptor (screen) only a single particle at a time. The receptor particles respond randomly, but only those on the lighter bands in the pattern will be influenced to respond at all. Therefore, if we leave the low light on long enough, the random spots on the screen must eventually build back to the same representation of the interference pattern.

In trying to decide which of the two slits the "photon" has passed through, any detector device, however subtle it is and wherever it is placed in the system, before or after the slits, must react with the system, if we are to take readings from it. The detector can then pass on a "photon", but the original wave function has already collapsed before it gets to the screen, therefore the intricate and delicate interference pattern disappears. If we set the experiment up with the detectors in place with without taking readings, we are not taking anything from the system, so we can still manage to get an interference pattern. But as soon as we take readings, the atoms of our monitoring system must interact with the light system, therefore the interference pattern will be disturbed and will disappear. There is no need for any communication through the system – that part of the waveform has collapsed in reacting with the detector, so can no longer interfere with the other part - so we see no interference pattern resulting from the two slits. Effectively, with the detector in place, emitting its own waveform after reacting with the initial waveform, there are now two systems, which will not interfere with each other.

The receptor particles will respond to the wave function according to the pattern of the field. As soon as there is a particle reaction at the receptor screen, the wave function collapses and the system is momentarily satisfied. The light source then delivers more wave functions. *Schrödinger's Wave Equation* effectively calculates the probability of the receptor response, not the photon position or velocity.

Heisenberg's Uncertainty Principle came about because we were looking to discern the position or velocity of something that *had* no independent position or velocity until we positioned something with which to make our measurements. The source atom had been excited, remaining in this state until it got a reaction from a receiver atom. What we must do to take measurements of such a "particle" is to see it through the reaction of a piece of matter – a measuring device. But then the particle *is* the measuring device. The electron or the photon only exists as a potential effect.

How does that effect come about? Any electron or photon, indeed all sub-atomic particles are the effect of a change in the energy levels of pure matter. In the case of the double-slit experiment, the matter particles of the light source are being excited, setting up a changing field in their surroundings that has been shaped by the two slits. The pure matter particles of the screen are responding to the effect of the excitation of the source. There are no independent particles in between.

Does this mean to say that photons and electrons are not particles? That question will be answered more fully in due course.

But the source will effect a change and the receptor will respond to that change – nothing whatsoever to do with the observer. Oh, yes, if you wish to make that response happen so that you can observe it, you will have collapsed the wave function – because you cannot observe the effect without making a receptor respond. But receptors respond to light all the time – matter is illuminated, whether we are looking at it or not. The observer is neither here nor there in what is happening. And there is a single response to a single waveform – no other worlds created, no alternative universes necessary!

*   *   *

In 1921, Albert Einstein received the Nobel prize in physics for explaining the photoelectric effect and for his contributions to theoretical physics. It was this explanation that first defined light as consisting of packets, or quanta (sometimes referred to as light being corpuscular in nature):

"When light is shone onto the surface of a piece of metal, a small current can be detected flowing through the metal. The light can be seen as giving energy to electrons in the atoms of the metal, making them move around, which produces the current. But not all colours of light affect metals in this way. No matter how bright a red light you have, it never produces a current in a metal, whereas even a very dim blue light will result in current flow. The problem was that these results couldn't be explained if light was thought of as a wave. Waves can have any amount of energy - big waves, lots of energy, small waves, very little. And if light is such a wave, then the brightness of the light affects the amount of energy - the brighter the light, the bigger the wave, the more energy it must have. The different colours of light are defined by their wavelength, which depicts the amount of energy they have. Blue light has shorter wavelength, red light longer wavelength, therefore blue light has more energy than red light, with yellow light somewhere in between the two. But if light is a wave, surely that must mean that a dim blue light would have the same amount of energy as a very bright red light. And if this is so, why won't a bright red light produce a current in a piece of metal in the same way as a dim blue light will?

Einstein's explanation of the photoelectric effect was to say that instead of being a wave, as was generally accepted, light was actually made up of lots of small packets of energy called photons that behaved like particles."

That, then, was the beginning of the misinterpretation of the nature of light. The definition of light assumed a particulate property which was spurious. The receptor responds to changes in the transmitter. A single photon is simply a single reaction in the receiver, responding to excitation from the transmitter. And the transmitter is an amalgam of single atoms, each one excited singly, each one producing its own energy wave. Every reaction at the receiver is in response to a single atom excitation at the source, the transmitter. If the light is red, each atom produces red light, which will effect no current flow at the receiver, the surface of the metal. Increasing the intensity light merely excites more single atoms, none of which will react with the metal. If the light source is blue, the receiver atoms in the metal will respond. Increasing the intensity of the blue light will increase the amount of individually excited atoms in the transmitter, increasing the number of stimulated atoms in the receiver, causing more current to flow in the metal.

The "quantum leap" excitation effect in the transmitter, the light source, instigates a similar effect in the receiver – but from particle to particle. So, throughout any such system, the photons and the electrons are never independently-existing particles, but are the "quantum-leap" responses of the matter (the receptor and the measurement instrumentation) in the field effect of the energy system, to the excitation of the source matter of the system.

What we have done, therefore, is distinguish between field-effects between matter particles and matter particles themselves. Photons and electrons are step-like field effects in waves, essentially part of the excited parent matter particles, instigating step-like responses from the receiving matter particles.

What, then, *are* matter particles? This, we shall examine in the next section.

Part Two

The Four Forces of Nature - The Questions

The four fundamental forces of nature are *The Strong Force, The Weak Force, Electromagnetism* and *Gravity*.

The first three have all been integrated into the standard model, with well-established calculations, confirmed by experiment and observation. So far so good. But it is the fourth of these, gravity, so ubiquitous in the real world, that refuses to be included in any model of quantum mechanics. Quantum gravity is seen to be the key to the discovery of any Grand Unifying Theory.

Gravity is the weakest of the four forces, but it has an infinite range. The electromagnetic force also has infinite range, but it is many times stronger than gravity. The weak and strong forces are effective only over a very short range and dominate only at the level of subatomic particles. Despite its name, the weak force is much stronger than gravity, but it is indeed the weakest of the other three. The strong force, as the name suggests, is the strongest of all four fundamental interactions.

Gravity is the force that presses us to the face of our planet, electromagnetism is the whole spectrum of light and magnetic field effects, the strong force is what holds the sub-atomic particles of atoms together and the weak force is what sometimes forces atoms apart (atomic decay, such as when uranium eventually decays into lead and other particles).

Of all four forces, it is gravity that stands alone, seemingly aloof and having nothing to do, that is no interaction, with the other three. But surely it must – gravity is everywhere, affecting everything. It makes no sense to have a theory of how everything works without gravity playing a major role.

The first to define the laws of gravity was, of course, Isaac Newton. But it was Albert Einstein who better described gravity. He devised two theories of relativity:

Special Relativity, published in 1905: concerned with the concept of time and its dependency upon the relative motion of observers.

General Relativity, published in 1916: concerned with the concept of acceleration, arguing that gravity is the geometric effect brought about by the curvature of space. Space is curved, Einstein postulates, due to the matter in it. The greater, or denser the matter, the greater the curvature of space (and therefore time).

It is the theory of General Relativity that cannot be incorporated into the standard model of quantum mechanics:

Quantum mechanics can be used to explain the chaotic and sometimes random actions within and between atoms. On this level, the effect of gravity is so insignificant that it is hugely overpowered by radiation and other forces. While general relativity describes an orderly and predictable universe at the macro level - Einstein was known to say "God does not play dice" - it is unable to explain the unpredictable subatomic environment that quantum mechanics describes. Conversely, quantum mechanics fails to explain the forces and relationships governing large objects, namely, gravity.

If both general relativity and quantum mechanics could ever be reconciled, with a "theory of everything", a single set of equations that could possibly explain all the mechanics of the universe. After Einstein formulated general relativity, he spent the rest of his life looking for such a theory. He never found it!

The Four Forces – The Solutions

Gravity, Electromagnetism, the Strong Force and the Weak Force – is it possible to define them and, in doing so, unify them?

The Nature of the Atom

All matter atoms consist of protons (with charge) and neutrons (without charge) in their nucleus. Each nucleus is depicted as being surrounded by as many electrons as there are protons.

Hydrogen is the simplest, lightest atom, with just a single proton in its nucleus. All the other elements are made up by adding protons and neutrons to the nucleus. The force that binds the particles together in the nucleus is the strong force. It acts on a very local level, but, as its name would imply, it is very powerful. It attracts protons to protons, neutrons to neutrons and protons to neutrons. So, what is the strong force, exactly?

Well, if we have two charged particles, two protons, they are attracted to one another by gravity, but they are also repelled from one another, because they both have similar charges, by the electrostatic force. Put two protons side by side and the electrostatic force is far, far greater than the force of gravity. It all works by these principles:

Gravitation

The *Universal Gravitation Equation* states the force of attraction between two objects, where the mass is considered concentrated at their centres of mass:

$$F = GMm/R^2$$

where

- **F** is the force of attraction between two objects
- **G** is the Universal Gravitational Constant, 6.67384×10^{-11} m3 kg^{-1} s^{-2}
- **M** and **m** are the masses of the two point objects
- **R** is the separation between the centres of the objects

Electrostatic

The electrostatic force equation is called *Coulomb's Law* and states the force of attraction between particles of opposite electrical charge. It also represents the force of repulsion for like charges:

$$F = k_e qQ/r^2$$

where

- **F** is the force of attraction or repulsion between two electrically charged particles
- **k$_e$** is the Coulomb force constant, 8.9875×10^9 N·m^2/C^2
- **q** and **Q** are point charges of the two particles
- **r** is the separation between the particles

The thing to notice about the above, is that both forces have a constant term involved in their calculations. If you were to put the figures into the equations, because the electrostatic force constant is so huge and because the gravitational force constant is so diminishingly tiny, the electrostatic repulsion is far more effective than the gravitational attraction. So, as the logic goes, it must be a stronger force than gravity at work in the nucleus of atoms – the strong force.

*　　*　　*

But what *are* these constants that play such a significant part in the calculations of these forces? The gravitational and electrostatic constants are not alone; constants occur frequently in nature. Take light, for example. The speed of light is a constant, at around 300,000,000 metres per second. It never alters, no matter how fast the measurer is travelling (in other words, it is not relative to the speed of anything else). But we must measure the speed of light *between* things; metres per second means space in time. Time and space are the same thing – *exactly* the same thing.

So, if we were to measure the speed of light between two particles, two protons, we would get 300,000 km/s, of course. But then what would happen if we were to get the two particles to exist in the same space and time? (This happens during nuclear fusion, which is what powers the sun.) It's easy to see that there could *be* no speed of light, as there are no metres and no seconds between them.

It was the German theoretical physicist Max Planck who suggested that the speed of light would be unified at distances approaching zero (1 planck length = 1.6162 x 10^{-35} metres – an unimaginably tiny distance!). At these minuscule distances, the constant speed of light is unified to 1, as are many other constants, including the gravitational and electrostatic constants.

If we were to go back now, to within the nucleus of the atom as it were, and do the gravitational and electrostatic calculations, unifying the constants to 1, we would find that gravitational attraction is by far the greater force, massively overpowering the electrostatic repulsion. Gravity is, therefore, very, very strong at these distances. In fact, when protons are fused, with only planck lengths or less between them, gravity *is* the strong force, increasing to infinity as we approach the atom's centre. Gravity becomes so powerful here, that nothing can escape it, not even light – which is the very definition of a black hole.

What we are suggesting then, is at the heart of every simple atom, and by that, we mean the hydrogen atom, the single proton, is a black hole.

*   *   *

Now imagine a black hole in space. Black holes, as we said, have a centre where gravity is so great that nothing can escape from it, not even light. So, just picture if this black hole were the only thing that existed in the universe – there would be no matter, just this pin-point of gravity surrounded by nothing. Could the black hole exist? Unless it has a gravitational effect on something, how can it be anything? There is no light, no matter – without them, there cannot be a black hole. There has to be something to have a hole in – if we don't, we have no black hole!

If we had a universe with one proton in it, there would be no spacetime – there would still be nothing. All matter is energy and energy must be gauged from some reference point, a datum. A single proton's energy measured from the datum of itself would give nothing. There must be a difference – there must be other matter.

So, let us then say we could have a universe with just two protons in it. Now we have distance between them, and we have energy levels to gauge from one to the other. Spacetime now exists between the two protons, so time passes. The distance between them doesn't just measure the time that has passed between them – it *is* the time that has passed between them! The time doesn't exist without this distance. And now we can have relative motion between the two particles. What we are saying is that particles only exist with absolute relativity to other particles.

And protons have a black hole at their nucleus: protons are gravity, then? They are essentially constructed of curved space, if we are to use Einstein's General Relativity, which we must. (General Relativity defines gravity as a geometric effect of acceleration, without the need for any "carrier particle", the graviton, which has been proposed, but never found.)

What, then, is curved space? Time (and therefore space) is continuously cascading into the black hole. This is gravity - there is essentially no difference between time cascading inwards and the black hole expanding outwards in all three dimensions in time. It is as if every atom is attempting to fill the whole of space and time. So, as the moon, for example, free-falls past the earth, the relative expansion of the earth and the moon take up the increase in space between them. The moon's velocity prevents the gap being "swallowed up" entirely. In this way, the moon experiences no force of gravity, but is locked onto a geometric pattern of orbit.

Drop two very different weights from the leaning tower of Pisa and (without air friction effects) they both stay suspended side by side and wait as the earth effectively takes up the space between them. Or imagine two astronauts suspended in space looking out towards the stars surrounding them at great distances. What neither space-traveller has noticed is the atmosphere-free planet behind them. They will feel nothing. They are not moving, apparently. But this thing, a whole planet is expanding behind them at an alarmingly accelerating rate. What would they ever know of its existence? There would be no clue that they are within the planet's "gravitational field". The planet will expand behind them and collect them onto its surface, holding them there, all apparently still, at rest - for the astronauts, probably for ever.

Matter creates space and therefore both matter and space are dynamic. What we have gauged as the nucleus of the atom is where we, i.e. other atoms, cannot easily go. As each atom expands outwards, it encounters other atoms all attempting to occupy the same space (and time). The black hole at the centre of each is attempting to consume everything around it. But it has encountered another black hole. Event horizon meets event horizon. Each particle is a continuum of curved space, but each has met a limit. As matter-objects ourselves, we can only ever gauge these limits and define them as a nucleus.

The matter particles will "fend off" one another unless they are forced into ever greater proximity. Given the right conditions, the right energy, the two particles can be forced to occupy the same space, the same time, in which they can operate as one – i.e. two protons become one proton and a neutron, which is what nuclear fusion is (not quite as simple as this, as four hydrogen atoms, protons, fuse to become two protons and two neutrons, the helium atom; but essentially, that is what is happening.) So, if we force two photons to operate as one unit, a different type of (composite) atom, there will be no distance between them, therefore no time exists between them. They are essentially the same thing.

So, again, what is it that binds the particles together so effectively, the so-called strong force?

Protons and neutrons are bound tightly together by gravitation rather than by some mysterious undefined force. However, as more atoms fuse, under hot, energetic enough conditions, the resultant element is heavier, its atoms more cumbersome, until their constituent protons have enough distance between them again to enable their repulsion and the element decays atomically (breaking apart). In this way, uranium, for example, will decay, shedding particles, to eventually become lead, whereupon it becomes stable.

Gravity, curved space is at the heart of every proton: it holds together the nuclei of heavier atoms and it battles with the repulsive charge forces (the black-hole event horizons) as the heaviest atoms decay.

Nuclear fusion is where two particles of curved space operate in the same space and time, and fission is caused by the charge effect of the more distant protons within the heavier atom structure, releasing a particle and a certain element of energy. This energy is the weak force – but it is not the *cause* of fission. And again, the energy released is the effect of the separated particles on other matter particles.

Gravity and the strong force are one and the same; the weak force is the interplay of gravity with the "event-horizon" forces, the repulsive charge forces of protons; and the electromagnetic forces (photons, electrons etc) are excitation effects.

Physicists created the new forces and particles – particles created by scientists do not exist! But then, which ones, out of the whole over-extensive list (see Appendix Four) do exist?

Part Three

The Different Masses of Particles - The Questions

When physicists investigate the nature of the different particles existing in the standard model of quantum mechanics, they are puzzled – it seems strange that some particles have different masses than others. Why strange? They must have, you might say, because they're different sizes. But then the photon, the particle of light, is said to be massless. It must be, to travel at the speed of light, because it *is* light. Nothing with mass could travel that fast.

How does this come about? Well, one theory is the Higgs Field, from which you will have heard of the Higgs boson: the Higgs field, made up of Higgs bosons, is everywhere, according to this theory, through which some particles can pass without any interaction, without being slowed down in other words, so that they are massless, able to travel at the speed of light, like photons; and some are made slower by interaction, hence made heavier, given mass, like electrons. Without the hidden Higgs field, in effect, everything would travel at the speed of light and nothing would have any mass at all.

What has happened is that the standard model has hit a problem, something that it cannot explain or describe, so invents a whole new field full of particles to fit a theory "tagged on" to the standard model. This has happened time and again, with new bits being added on to try to make the standard model work. A new theoretic attachment almost always requires a whole new set of particles and so, certainly in the case of the Higgs boson, the people at CERN* start searching.

And, in this case, finding ... haven't they?

*(At CERN, the European Organization for Nuclear Research, physicists and engineers are using the world's largest and most complex scientific instruments to study the basic constituents of matter - the fundamental particles. The particles are made to collide together at close to the speed of light. The process gives the physicists clues about how the particles interact, which supposedly provides insights into the fundamental laws of nature.)

The Different Masses of Particles - The Solutions

Who Ordered That? – a plethora of false particles!

By the mid-1930s, physicists believed they had identified all the subatomic particles of nature – the proton, neutron, and electron of the atom. But in 1936 the muon (see Appendix Two) was discovered – a new particle with such surprising properties that Nobel laureate I.I. Rabi quipped, "who ordered that?" when he was informed of the discovery.

From then on, many new "particles" have been discovered, many more suggested and looked for – like the graviton, for example: ubiquitous, yet undetectable. And others, like the Higgs Boson, which supposedly gives some particles mass and not others - as with electrons, which have mass, and photons, which do not. Let us concentrate, then, on those two sub-atomic particles, the electron and the photon.

*   *   *

When as electron moves from one energy level (shell) within an atom to a lower level, a photon is released. The photon is massless and is travelling at the speed of light. And yet, when another atom responds to the photon, mass (energy) is transferred. So how could the photon be massless? And if it isn't, how could it travel at the speed of light, when something carrying mass from one place to another cannot travel at the speed of light?

And then the electron has mass, so travels at a speed some way below the speed of light. What are they, these sub-atomic particles?

Well we have already ascertained that they are effects of the changes in matter particles, in protons and neutrons. Let's take the photon first:

The atom decreases its energy, climbing down from one level to another, and there it sits at the lower level. The effect is everywhere around it, as it has no actual boundary unless gauged from another atom. So everywhere around the altered atom is instantly and simultaneously altered – in other words, at the speed of light.

Now look at the electron. The atom is suddenly excited, one may say over-excited, it will then "shuffle off" the extra energy. The effect on its surroundings is a "peak" of energy effect travelling outwards in all directions. The energy wave propagates in all directions at once, travelling at a specific speed (less than the speed of light), carrying energy outwards. But this effect must never be mistaken for a particle, in any way distinct from its parent atom. Until it reacts with another atom, the travelling wave is an integral part of the parent atom, which has no boundary. The wave effect is still centred on the black-hole nucleus of the atom, its energy still an integral part of the energy system of that atom: a condition which remains until it reacts with that other atom.

So, the photon and the electron can be viewed as the surrounding effect of the atom having changed size, in the case of the photon, and having changed shape, in the case of the electron.

Then, from the above descriptions, the photon and the electron can therefore be looked upon as a standing wave and a travelling wave, respectively:

Standing Waves

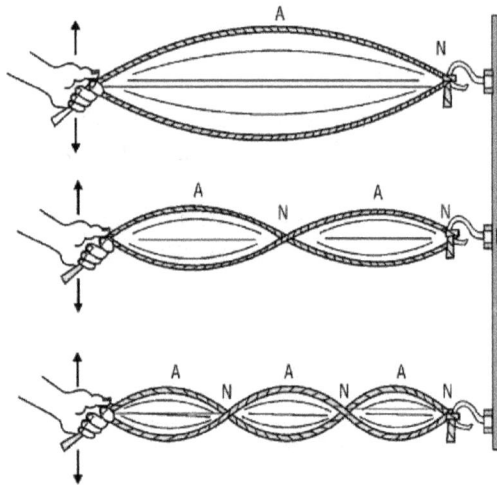

Consider the string set-up as depicted above. There you can see different wavelengths, relating to the level of excitation, but the effect is all the way along the string, all the time. This is the same effect as a photon. By varying the frequency so that the pulses are produced at certain intervals, there can be produced fixed points of destructive interference (nodes) and fixed points of constructive interference (antinodes).

Travelling Waves

Travelling waves, on the other hand, move from place to place, transporting energy, in the same way as the electron does. Travelling waves can have any frequency, just like standing waves. The travelling wave is like a seaside wave that can knock you off your feet with its shifting energy.

It must be realised that standing waves or travelling waves can move in one, two or three dimensions.

*   *   *

But neither the photon nor the electron can exist without their relationship to the originating atom and then their effect on another atom. Until the receiving atom reacts to the photon or the electron, the photon or electron is still part of the energy system of the mother atom.

Do photons and electrons exist as independent particles? No. They are part of the source-atom until they transfer to the receiver-atom as effects. (Although there have been electrons supposedly isolated and kept for weeks by physicists, with measurements made of their attributes – for the discussion of this, see Appendix Six.)

If photons and electrons have no independent existence, then what about all the other sub-atomic particles?

Well, we can see how the photon as a standing wave is instantaneously everywhere, i.e. it has no mass, when the electron is a travelling wave of energy propagating outwards from the mother atom – therefore, we require no Higgs Field to attribute some particles their mass. Protons and neutrons have mass simply because they are sheer energy. But there is no Higgs boson.

So, what have the physicists and engineers at CERN been discovering, with their new particle that seems to suggest the Higgs boson?

Let's just step back and look again at what we've done with photons and electrons. They are reactions to changes, causes and effects of changes of energy levels within protons and neutrons. So, if we smash particles one into the other, we will see effects. And as we continue increasing the energy of the accelerations leading up to the collisions, we will see more and more energetic effects.

But do engineers at CERN measure the accelerated particles themselves? No. They must use measuring instruments. In other words, they are measuring the effects of these dramatic collisions indirectly, by their effects on unaccelerated particles.

The disturbances in the fields surrounding atoms will always have effects on other atoms. That is what we're seeing. Most of the detected "sub-atomic particles" have a tiny, almost immeasurable lifespan – in other words they are the ephemeral effects in the overall construction of the fields between atoms, the very structures of the protons and neutrons. Even the quarks are glimpsed disturbances that disappear as the atoms regain their equilibrium state. The different detected quarks are simply the different possible effects of such collisions on the equilibrium state of the matter in the measuring instruments.

All the effects are measured by instruments that are collecting the temporary disturbances. These disturbances appear to be particles, in the same way as photons and electrons appear to be particles. Feynman diagrams (see Appendix Five) are a clever way of mapping possible field disturbances and interactions by identifying combinations of effects – but all these effects are only ever realised by the reactions of equilibrium-state protons and neutrons.

And, during nuclear fusion, when two protons are fused with another two protons they will form an alpha or helium particle. Two of the protons have effectively turned into neutrons. Release a neutron and it will turn back into a proton. The neutron is, therefore, an absorbed proton.

So, what else is there?

Nothing. Protons.

Only protons.

Part Four

Quantum Entanglement - The Questions

This effect is, as Albert Einstein called it, truly spooky action at a distance. Under certain circumstances, like when a laser beam is fired through a crystal of some particular type, a pair of photons, or light particles, can be produced together. That is to say that they are separate effects, but are somehow intertwined, truly *entangled*.

Now, when photons or electrons exist together in this way, the *Pauli exclusion principle stipulates* that no two identical particles may exist at the same energy level. In other words, there must be a difference between the two.

*   *   *

The Pauli exclusion principle, (devised by Austrian physicist Wolfgang Pauli in 1925), originally applied to electrons in an atom. It states that no two electrons in any atom can have the identical quantum mechanical state. The principle has been extended to any two particles existing together anywhere, such as two entangled photons. So these photons must either be at different energy levels, or they must differ with respect to an energy property called spin (although photons do not actually spin – which makes spin such a confusing description. Better to regard spin as an energy attribute, and leave it at that, for now).

According to the principle, the two particles, as they share the same energy level, cannot be identical - the two particles are different in their spin, one being the opposite of the other.

*   *   *

Now, the interesting thing about all this, as far as quantum entanglement is concerned, is when we examine these entangled photons; if we look at one and it has spin in one direction, the other has spin in the opposite direction, which does not violate the exclusion principle. But what we can do with this pair of photons is to separate them, across billions of metres, potentially. When we do this, examining one will tell us the quantum state of the other.

Fine, you may say, as Einstein did, but the two are one thing or the other when they are formed, one at one spin, one at the other, like a pair of gloves. But other physicists were suggesting that particles do not have any quantum state until we measure them. So, which was it?

Well, along came a fellow called John Stewart Bell in 1964. He devised a system incorporating what is called "Bell's inequalities", which managed to demonstrate that the particles had no fixed quantum state until they were examined. Einstein was wrong. (It is not essential to understand how Bell's inequalities proved this – but there is plenty of literature available on the subject, should you wish to research this further.)

The essential thing about what Bell did with his inequalities was to show that the two entangled particles were not already "right-handed and left-handed" on production, as Einstein suggested, but when one was fixed into one kind of spin in one place, by being examined, then the other took the opposite spin. This meant that when one did one thing, the other one somehow knew about it and did the opposite thing. Instantaneously. And this can happen over millions of miles, potentially. So how can they communicate instantaneously over such vast expanses, when, according to Einstein's Theory of Special Relativity, nothing can travel faster than the speed of light?

Quantum Entanglement - The Solutions

Now let us say that there are not two entangled photons at all, but a polarized form of a single waveform - but, as always, still part of the mother atom energy system. Polarization means that the waveform has been broken down in this way:

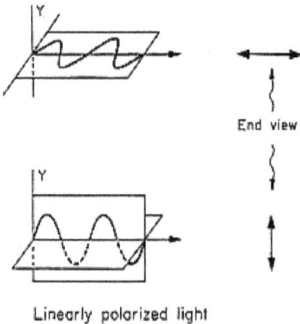

End view

Linearly polarized light

These polarized waveforms propagate outwards together, in all three dimensions in time, until, as usual, there is a reaction from another atom. In one place, an atom will react to one polar form, which then collapses, leaving the other form to react with another atom – the distance between the two events is irrelevant – the two reactive atoms are responding to the different aspects of the same thing: the mother atom. Therefore, the spin-effect of the reaction has not been decided upon until the first reaction, leaving only the opposite spin reaction available to the second reactive atom. In this, there is no need for any kind of communication across the system.

The confusion in understanding what is happening with quantum entanglement perfectly illustrates the error that has led to the apparent impossibility of integrating the micro and the macro worlds: sub-atomic "particles" have been given independent particulate properties, when they are always either part of the mother atom itself, or else powerful but ephemeral disturbances in the equilibrium state of matter, brought about by violent atomic collisions but measured on equilibrium-state instruments.

The above error always leads to an apparent "communication" between "particles", when all that is happening is simple atom-to-atom reactions.

Part Five

Matter and Antimatter - The Questions

When the universe came into being at the big bang event, matter was split from its counterpart antimatter. In energy terms, antimatter is the opposite to matter. This means that, should the two meet, they counteract and cancel each other out. The sum energy of the universe is, as it has always been, zero.

But as scientists use more and more advanced and sophisticated instruments and probes to explore and measure the universe, it has become clear that there is practically no antimatter.

What has happened to it all? There should be as much antimatter as there is matter; not roughly as much, but exactly as much!

So, where has all the antimatter gone?

Matter and Antimatter – The Solutions

So, where is all the antimatter? Why didn't the matter and antimatter annihilate one another during the big bang?

Here's the answer: matter is curved space; and time and space are the same thing. Without matter, there is no space or time. Matter creates time. Antimatter, therefore, creates anti-time – time that runs in the opposite direction, that is. That is not my idea – it has been stated many times, by many theorists. For example, the Feynman-Stueckelberg Interpretation (Richard P. Feynman and Ernst Stueckelberg) states that antimatter is identical to matter but moves backwards in time.

What it means though, is that as soon as matter and antimatter were created in the big bang beginning of our universe, they existed at different times. They both existed in each other's past, matter moving away into its future, antimatter moving into *its* future. Matter and antimatter could not annihilate one another, as, at the very first moment of time, they existed at different times. They could never contact one another.

This means to say that the universe is older than itself. Every second that goes by here, so our past expands into the past. We cannot see the antimatter as it is just too far away. We'd need to see before the big bang; and that we shall never be able to do. But this shows the nature of infinity, that it is as dynamic as everything else in the universe. What was before the big bang? The universe was – our antimatter universe, that is. And could there be life in the antimatter universe? Yes. The antimatter universe is essentially exactly like our own – because it *is* our own universe! The only difference is, that from there, we are antimatter, with time on this side expanding into their past.

There is a symmetry to the universe, simple, beautiful and wondrously balanced, with no alternatives, no branes (membrane universes) or hidden dimensions. They are simply not necessary.

N.B. This book, and this theory, was first published on 1st August 2013. Since then, a small team of scientists, Dr Julian Barbour of College Farm in the UK, Dr Tim Koslowski of the University of New Brunswick in Canada and Dr Flavio Mercati of the Perimeter Institute for Theoretical Physics, also in Canada, have published a scientific paper which reaches the same conclusion.

This paper was published in Physics, in the Physical Review Letters, on 31st October 2014 (DOI: 10.1103/PhysRevLett.113.181101), "Identification of a Gravitational Arrow of Time".

A simple overview of this paper can be read in the Daily Mail on-line from 10th December 2014, using the following link:

http://www.dailymail.co.uk/sciencetech/article-2868238/Did-Big-Bang-create-mirror-universe-time-moves-BACKWARDS-New-theory-explain-past-future.html

Before 1st August 2013, there was no other published information on this theory. It was published here first.

Part Six

The Questions - Summary

So, we have posed the questions and we have proposed the solutions using the Theory of Unified Absolute Relativity. We have:

1) Fully explained the effects of the Double Slit Experiment by determining the effect that gives matter its wave/particle duality; and resolved the enigma by a simple but effective new description of light.

2) Incorporated gravity into a quantum model (by determining the nature of matter particles).

3) Adequately described and demystified the process of quantum entanglement.

4) Described why particles have different masses and why some (photons in particular) are apparently massless.

5) Determined what has happened to *all* of the antimatter in the universe.

Now show me a quark. The best that can be done is to see the effect the "quark" disturbance has had on another atom.

Shall I show you an atom? Here's a droplet of water. Hydrogen atoms. Oxygen atoms. See them? Feel them, in contact with your skin.

The "quarks" are glimpsed through powerful and violent collisions – the different kinds of "quark" denote the possible atomic responses to the distortions in the equilibrium-state of all the other atoms. No atoms are destroyed or altered in this process. The kinetic energy of the accelerated particles is almost instantly redistributed as other types of radiation by other atoms until equilibrium is restored. There were no "quarks", just as there are no "quarks" at the nucleus of atoms.

Here's a New Certainty Principle – slap the palm of your hand down onto the table-top; you can be certain that the table, as well as your hand, is there. The table, and everything we are as integral parts of the universe, is made of the elements. The elements are listed in the periodic table. (There are plenty of good examples freely available on the internet). You will see that all matter comprises of protons and their attendant neutrons (absorbed protons). With this in mind, I would determine that protons and neutrons are not sub-atomic particles, as they are often categorized, but the essence of all particles. They are the real particles, the ones that exist in actuality and in perpetuity. All atoms are either hydrogen, the simplest atom, or composites.

The New Certainty Principle ascertains that you can be sure where the protons and neutrons, where the atoms are when the atoms of your hand move through the atoms of the air and strike the atoms of the table. The Heisenberg Uncertainty Principle applies to "sub-atomic" particles, which aren't particles at all, but potential effects on real particles (atoms). So, when an atom responds to a photon-effect, you can be sure where that effect is taking place because you can see it (or you see the effect of many combined atoms when matter is illuminated). Sub-atomic "particles" are never divorced from their parent atom: that is the essence of what we have established. In fact, these effects are atom to atom responses, even when the atomic field effect has been polarized to affect two different atoms in two entirely separate places. By this, we have explained what is happening in Quantum Entanglement, as well as in the apparent illogical behaviour illustrated by the Double-Slit Experiment. By accepting that atoms are curved space, that all the other so-called "particles" are ephemeral field effects, we have established that protons are the only fundamental particle there is – and protons are black holes, units of curved space, sheer gravity. Gravity alone is at the heart of every atom, holding together the heavier elements and causing all the other field effects as real particles exchange energies.

In the end, the one natural force that refused to be incorporated into the standard model of quantum mechanics has been shown to be the fundamental force driving all the other effects. The universe and everything in it is made of gravity.

Gravity is energy, a black hole extending outwards indefinitely – but the energy segregated on the other side of the big bang is negative to the energy on this side. Matter and antimatter remain in perfect equilibrium throughout the universe, the one and only matter and antimatter universe, here.

Gravity is what causes it all.

Gravity is all we are. It is everything.

Just gravity.

Appendix One

Why is an atom not always susceptible to a photon?

Atoms respond to changes in atoms. Look again at the depiction of a standing wave, which is analogous to a photon:

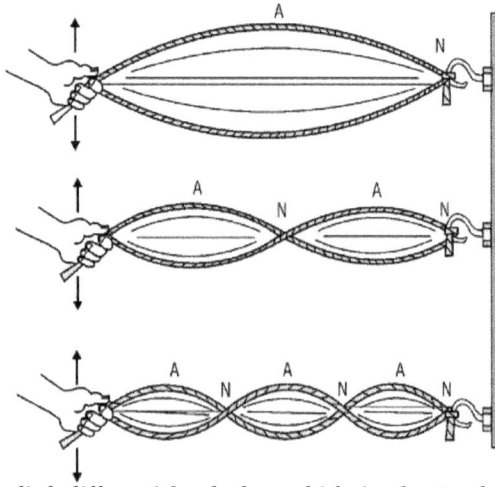

If you have studied differential calculus, which is about calculating rates of change, you'll understand that the rate of change of the waveform of the string at its maximum or minimum is zero. If you haven't studied calculus, look at it like a "big-dipper" ride – at the top or the bottom you experience the effect of going from rising to falling, or falling to rising respectively. So, going from up to down, there must be a moment of nil effect. But if you are, say, plunging downwards, you will soon experience a maximum rate of change, a maximum acceleration. It is this change, this acceleration that elicits atomic responses.

So, if the receiver atom is so positioned that it encounters the photon-effect from the matter atom at or near the top or the bottom, it senses no change. So, no response. If it is in a position to encounter the photon-effect half-way down or half-way up, i.e. at maximum rate of change, it will be sensitive to the change and may respond. All in all, there is probably as much chance of the encounter occurring at a point of zero or too little change with no response, as there is of the encounter occurring at a point of sufficient change to effect a response.

Appendix Two

The Graviton and The Muon

The Graviton

According to the standard model of quantum mechanics, all forces have carrier particles, including gravity: the elusive graviton.

But this approach completely ignores Einstein's definition of gravity as curved space, a concept which requires no other particles, nothing but the mass of the gravitational object itself. Protons and neutrons create curved space. They a black-hole pin-points of gravity expanding outwards, curving the structures of space and time. Therefore, the proton *is* the graviton. Protons are everything.

The Muon

The muon is like an electron, but has about 200 times the electron's mass. They are raining down on us all the time, because they originate from cosmic rays, that is protons ejected by stars, colliding with the atoms of our atmosphere. They have a short life; around 2 millionths of a second.

This mean, of course, that as accelerated protons from space are colliding with the atoms of the earth's atmosphere, as in any highly energetic collision, there is a violent reaction as the kinetic energy is redistributed. Accelerated particles colliding with other particles will produce different effects according to the energies, the types of particle and the surroundings. Muons are like every other "sub-atomic particle" produced in his way – they are temporary effects.

Now physicists suspect that the muon may be capable of some kind of new type of interaction, in addition to the gravitational, electrostatic, strong and weak interactions – in other words, another force to add to the natural four. It is hypothesised that this fifth force could have been responsible for annihilating most of the antimatter following the big bang event at the beginning of the universe.

So now there will be an attempt to complicate the standard model still further with another force, carried by over-fed electrons with a lifespan of 2 millionths of a second. It just has to be more simple than that! Please see *Part Five – Matter and Antimatter*, for a logical and far more satisfying solution.

Appendix Three

"According to string theory, the universe is made up of tiny strings whose resonant pattern of vibration are the microscopic origin of particle masses and force charges. String theory also requires extra space dimensions that must be curled up to a very small size to be consistent with our never having seen them. But a tiny string can probe a tiny space. As a string moves about, oscillating as it travels, the geometric form of the extra dimensions plays a critical role in determining resonant patterns of vibration. Because the patterns of string vibrations appear to us as the masses and charges of the elementary particles, we conclude that these fundamental properties of the universe are determined, in large measure, by the geometrical size and shape of the extra dimensions. That's one of the most far-reaching insights of string theory." - Brian Greene, *The Elegant Universe*

It was reading Mr Greene's "*The Elegant Universe*" that finally convinced me that string theory, in any of its subsequent and increasingly fantastic and convoluted guises, will never be able to offer anything like a plausible and coherent theory of everything. With all its extra dimensions, curled up really, really, really, really small, so no one will ever be able to detect or position them, and all its branes, that is membranous universes set out in – in what, I've never been able to ascertain – string theory has become a playground for serious scientists to give vent to their sci-fi fantasies, chasing childish dreams that are bigger than our universe itself.

*　　　*　　　*

Over the years, string theory has appeared in many guises, usually growing in convoluted complexity as they go. Here are just some of them:

Bosonic string theory - Superstring theory (or Supersymmetric string theory) - Type I, Type IIA, Type IIB, Heterotic string theories (Type HE, Type HO) – M-theory - Matrix theory - Brane world scenarios - Randall-Sundrum models (or RS1 and RS2) – F-theory – Etc.....

*　　　*　　　*

String theory does this, time and again: it proposes a bizarre event, like rips and tears in time, then it proceeds to search for some convoluted means by which to mathematically calculate what might happen to a speculative particle (or sparticle!) under those speculative conditions. Then, whenever the mathematics comes out to somewhere around some predicted level, Mr Greene and his colleagues call this a "breakthrough". All this, without a single testable experiment upon which to base the original hypothesis.

You do not need to be a scientist to see that string theory is based on bad science: it is a cerebral exercise in abstract theory and mathematics, intellectually demanding, of that there is no doubt; but descriptively devoid of any significant insights into the interactions of the natural world.

Appendix Four

List of Particles

Out of interest, but also out of a sense of personal exasperation, I have included below a list of many, but not all the particles in the current standard model of particle physics. But please bear in mind as you look down this list that what we must now do is to distinguish between actual particles and their effects on other particles. For example, W and Z bosons, the list says, are the carrier particles of the weak force, supposedly responsible for radioactivity. But radioactivity, as complex atoms decay into simpler ones, is the release of energy as either neutrons give up some mass and decay into protons inside the atom, or the atom splits up because of the conflicting forces of gravitational and static field effects of the protons within the complex atoms.

Then there are the quarks; as we have determined, curved space, gravity is at the heart of every atom, not quarks.

From there, look at the group Fermions – all matter is made of fermions. It means that fermions are made of quarks, either two or three. Protons are supposedly made of three quarks and protons never decay. They are completely stable, as far as we can tell, for all time. But the other fermions, the ones that consist of two quarks? Well, take this one example:

The Kaon (containing a "strange" quark and an "up" quark) has a mean lifetime of 1.2384×10^{-8} seconds (around one, one-hundredth of a millionth of a second). So, like the quark itself, the kaon is detected under highly accelerated conditions which include violent collisions, where the kinetic energies of excited particle systems are detected, momentarily, under extremely unrealistic test conditions.

Show me a quark, a kaon, a photon or an electron – all that is ever shown is the effect of these excitations on the atoms of matter. These effects, these responses depend upon the nature of the stimulation from the field system of the parent (excited) atom. If you want to see an atom, just look. They are all you *can* see. And you are seeing them purely because atoms are being excited by light, by the changing fields of other excited atoms. The atoms of your eyes are being similarly excited by those excited atoms, creating electrical pulses to your brain. Atoms exciting other atoms.

In addition, as you go down the list, there is the graviton, which does not exist. And then there is the Higgs boson, which does not exist.

All there is, all there will ever be, is atoms, with protons and neutrons at their heart. Protons are eternal. Neutrons are absorbed protons; released, they will soon decay into protons, that is they will give up a bit of their energy to the equilibrium state (the ground state) of all the atoms surrounding them.

*   *   *

The List

All particles are either fermions or bosons.

Fermions. (half-integer spin 1/2, 3/2, 5/2, etc.) Matter is made of fermions. Fermions obey the exclusion principle; they cannot be in the same place at the same time.

Bosons. (integer spin 0, 1, 2, etc.) Forces are carried by bosons with non-zero spin. Bosons do not obey the exclusion principle; they can pass right through each other.

Elementary particles. Elementary particles are not composed of other particles. The elementary fermions are the quarks and leptons. The elementary bosons are the photon, W and Z bosons, gluon, graviton, and Higgs.

Quarks. (spin 1/2) The protons and neutrons in the nucleus of an atom are made of quarks. There are six types or "flavours" or quarks: down, up, strange, charm, bottom, and top. Each comes in three "colour" charges: red, green, and blue.

Leptons. (spin 1/2) The six leptons are the electron and its two heavier sisters, the muon and tau, and the three lightweight neutrinos, the electron neutrino, muon neutrino, and tau neutrino.

Graviton. (spin 2) Gravitons [predicted] carry the gravity force.

Gluon. (spin 1) Gluons carry the strong force, also called the nuclear force or colour force. The strong force holds quarks together.

W± and Z bosons. (spin 1) W± and Z bosons carry the weak force. The weak force is responsible for radioactivity.

Photon. (spin 1) Photons carry the electromagnetic force. Photons are particles of light. Light is an electromagnetic wave.

Higgs. (spin 0) The Higgs boson [predicted] is an excitation the Higgs field. The Higgs field gives other particles their inertial mass.

Electroweak W and B bosons. (spin 1) W1, W2, W3, and B bosons carry the electroweak force. When the electroweak force split into the electromagnetic and weak forces, the W1, W2, W3, B, and Higgs remixed to make W±, Z, photon, and Higgs.

Composite particles. Composite particles (hadrons) are composed of other particles. The main types of composite particles are the baryons and the mesons.

Baryons. (spin 1/2, 3/2) Baryons are fermions composed of three quarks. The most important baryons are the two nucleons: the proton (up-up-down quarks) and the neutron (up-down-down quarks). Some other baryons are the sigma, lambda, xi, delta, and omega-minus.

Mesons. (spin 0, 1) Mesons are bosons composed of a quark and antiquark. Some mesons are the pion, kaon, eta, rho, omega, and phi.

Antiparticles. All particles have a corresponding anti-particle that is identical in many ways but opposite in others; for example, the mass and spin are the same but the charge is opposite. An uncharged particle may be its own anti-particle.

*   *   *

Hypothetical Particles

All new theories seem to demand new particles, so there are many, many of them. Here are just some:

Strings. String theory postulates that all elementary particles are really tiny strings with different vibration modes.

Sterile neutrino. A particle that has no interactions except gravity.

Graviton variations. The spin 1 graviphoton and the spin 0 graviscalar (also known as the radion or dilaton).

Axion. A particle proposed to explain the absence of an electrical dipole moment for the neutron. Supersymmetry adds the axino and saxion.

Goldstone boson. A type of spin 0 particle that that is necessary wherever there is a broken symmetry. Supersymmetry adds the goldstino and sgoldstino. The majoron is a type of Goldstone boson.

X and Y bosons. Particles mediating a grand unified force, analogous to the W and Z bosons.

Mirror particles. Particles with left/right (parity) opposite of known particles.

Magnetic monopoles. North and south monopoles, analogous to positive and negative charged particles. Some theories also propose a magnetic photon.

Tachyon. A particle that travels faster than light and backward in time.

Exotic baryons. Fermions composed of three quarks plus other particles. The pentaquark has five quarks.

Exotic mesons. Bosons composed of other particles, but not merely two quarks. The tetraquark has four quarks. The glueball is composed of gluons.

Appendix Five

Feynman Diagrams and Virtual Particles

Feynman Diagrams

Being a great admirer of Richard P. Feynman, having learned a lot from his writings (*The Character of Physical Law*, in particular) and from his recorded lectures and talks, I wanted to include something about Feynman diagrams. Ingenious and intriguing, these pictures brilliantly represent interactions between particles:

Particles entering or leaving a Feynman diagram correspond to real particles, while intermediate lines represent virtual particles.

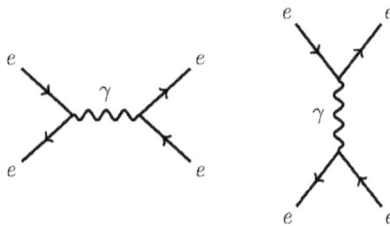

The diagram illustrated above represents the interaction of two electrons. Each electron is represented by a straight line, which exchange a (virtual) photon and then repel one other.

In reading diagrams like the one featured above, it is worth remembering that the particles featured are not divorced from their parent atoms, but these interactions are the effects of interactions *between* the atoms concerned, never free electrons or photons existing in their own right.

*   *   *

Virtual Particles: What are they?

Anything labelled "virtual" tells us that it is a construct, that it does not exist in the real world. And that doesn't just mean that it doesn't exist without its parent particle, like a photon or an electron; but that it does not exist at all! It is a description of the interchange between energy fields. The Feynman diagrams above show electrons, which are integral to their individual protons, with the virtual particles, in this case virtual photons, merely helping to describe the reaction between the two mother protons.

Appendix Six

Spin and the Pauli Exclusion Principle

The next four paragraphs, in italics, all relate to the atomic model formulated by Danish physicist Neils Henrik David Bohr (7 October 1885 – 18 November 1962). The Bohr model depicts an atomic nucleus of protons and neutrons orbited by attendant electrons.

Perhaps the easiest way of thinking about the property of particle spin is as a rotation of particles around their own axis. Spin obeys the same mathematical laws of angular momentum as do spinning objects in classical physics, such as the Earth, with really only two important aspects to consider: the speed of rotation and the direction of the axis it rotates about (referred to as "up" and "down").

This property of spin leads us to the Pauli exclusion principle (formulated by Wolfgang Pauli in 1925). This states that no two identical fermions (electrons, for instance), may occupy the same quantum state simultaneously (although two electrons, for example, may acquire opposite spin in order to differentiate their quantum states).

To understand why, it is necessary to know that, according to the Bohr model of the atom, electrons in an atom (which exist in the same quantity as the number of protons in the nucleus of the particular atom) are constrained to occupy certain discrete orbital positions or "shells" around the nucleus. The closer electrons are to the nucleus, according to this model, the more strongly they are being pulled in and the more energy would be required to free them from the nucleus. The innermost shell can accommodate just two electrons, one with spin "up" and one with spin "down" in order to differentiate their quantum states. The next shell out, in a higher energy level, can accommodate a further eight, the next a further eighteen and the next thirty-two.

It is the Pauli exclusion principle which dictates this arrangement and effectively forces electrons to "take up space" in the atom through this arrangement of shells. By recognizing that no two electrons may occupy the same quantum state simultaneously, it effectively stops electrons from "piling up" on top of each other.

The Pauli exclusion principle is one of the most important principles in quantum physics, largely because the three types of particles from which ordinary matter is made (electrons, protons and neutrons) are all subject to it, so that all material particles exhibit space-occupying behaviour. Interestingly, though, the principle is not enforced by any physical force understood by mainstream science. When an electron enters an ion, it somehow mysteriously seems to "know" the quantum numbers of the electrons which are already there, and therefore which atomic orbitals it may enter, and which it may not.

If you were to look up the Periodic Table, you'd see that the diagram, below, shows an atom of the element Caesium, with 55 protons. The Bohr model diagram shows the 55 protons, with 78 neutrons in the nucleus, while 55 electrons float about in their respective shells outside the nucleus. All the elements, all the stuff we are, is made up like this, with protons and neutrons stuck together, while electrons apparently orbit this arrangement.

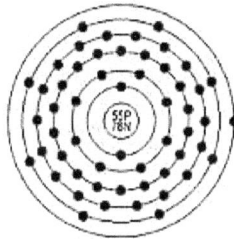

So, atoms are made up of protons and neutrons. The neutrons are absorbed protons and are electrically neutral, (hence the name) and have no charge. Only the protons are at an energy level at which they will repel and be repelled by other protons, like two negative-end, like-sign magnets. Physicists have determined this energy attribute as an individual particle, the electron, which then serves to neutralise the charge of the protons, and which somehow orbits the nucleus, but which doesn't orbit the nucleus, but has angular momentum, but is just a probability cloud – all these things at once.

Unified Absolute Relativity of course simply shows that one energetic proton wants to repel the other as their black-hole event horizons meet and refuse to "marry". No electrons at all. Protons are forced to "marry" during the fusion process, fitting one to the other by a combination of protons and neutrons (absorbed protons), nestling together according to their energy attributes.

The protons themselves have the energy attributes that combine in particular ways to form more complex atoms. If a proton has, say, horizontal spin, then vertical spin is available. So, it is that the spin is with the proton itself, never any non-existent electron. The spin of the proton affects its ability to affect or be affected by other protons. Look at the above – the protons in the middle of the complex atom are bound more firmly, providing the inner shells, and as we go further out, the outer atoms provide the outer, less firmly bound shells.

When an atom is "ionised", it apparently gains an extra electron. In the standard model, a free electron approaches and nestles into the above arrangement, but according to the Pauli exclusion principle outlined above, the electron must somehow "know" the spin it must have to fit into the shell structure of the atom. This it does, taking the spin of a vacant space in the shell structure, ensuring that no two electrons sit at the same energy level.

But Unified Absolute Relativity sees it another way: The atom is affected by a separate, excited atom that either has the excitation attribute, the correct spin, to influence the receiving atom (taking positive energy from it), or it does not. The spin is therefore with the parent and the receiving atoms, not with bogus electrons.

Again, giving the electron particle position and attributes of its own introduces spurious and curious "communications" problems that will bemuse everyone, but which the physicist will simply shrug at and declare once again that the quantum world cannot be described in classical terms. But, by this, we have proved, once again, it is the spurious particulate nature of sub-atomic particles that has denied any classical description, not the nature of Nature herself.

Appendix Seven

On Physicists' Attempts to Capture an Electron

Physicists have conducted experiments employing a novel technique in which a single electron is confined for weeks at a time in a "trap" formed out of electric and magnetic fields. In effect the electron and the confining apparatus make up an atom with macroscopic dimensions and an extraordinarily massive nucleus.

The trap for the electron consists of a special configuration of electric and magnetic fields. The electron occupies the central cavity formed by the two cap electrodes and the ring electrode, which are machined to a mathematically determined shape: they are hyperboloids Two additional electrodes, the guard rings, compensate for imperfections in the electric field. The entire apparatus, which is about an inch and a half in diameter, is immersed in liquid helium and inserted into the core of a superconducting magnet the electron is bound by the combination of static electric and magnetic fields in the trap, much as an electron in an atom is bound to the nucleus. Here the part of the nucleus is played by the apparatus, or even by the earth, on which the apparatus rests, and so the atom is called geonium, the earth atom.

But the above description is surely that of the model of a proton, very cleverly sculpted out of carefully constructed force fields. The electron described as having been "captured" is similar to the charge effect in a real proton – in other words, the "electron" is just as much an integral part of this modelling system as it is a part of an actual proton. This electron, like all others, as I have said time and again in this book, does not exist without its mother atom. And what we have, in the description above, is a model of an atom, not a captured electron.

Appendix Eight

One Last Question Posed – Unanswered

Stars produce some elements by fusing lighter elements – i.e. hydrogen fusion to produce helium, helium fusion to produce lithium (not quite as simply as this, but essentially that's what happens). But there is insufficient energy available in any star process to produce the heavier elements. For these, we have to thank the violence of the end of the life of the stars, whereby more energy-dependent fusions produce all the heavy elements which are then gathered together to form new solar systems such as our own.

My question is this:

Why did none of these fusion processes occur during and just after big bang, when all the simple elements in the universe were close enough and therefore energetic enough to fuse with one another? How did hydrogen survive this process at all? Why isn't the universe made of uranium, or gold, or iron?

*   *   *

Speculations on Dark Matter and Dark Energy

Dark Matter: The source of extra gravity

Astronomers have observed that the gravitational effects in our universe don't match the amount of matter that can be seen. To account for these differences, physicists have suggested that the universe contains some kind of mysterious form of matter we can't observe, dark matter. The suggestion is that there is around six times as much dark matter as normal visible matter.

In the 1930s, Swiss astronomer Fritz Zwicky first observed that some galaxies were spinning so fast that the stars in them should fly away from each other. In 1962, astronomer Vera Rubin made the same discoveries and managed to maintain her focus on the problem; by 1978 she had studied 11 spiral galaxies, all of which (including our own Milky Way) were spinning so fast that the laws of physics said they should fly apart. Together with work from others, this was enough to convince the astronomy community that something strange was happening.

Whatever is holding these galaxies together, observations now indicate that there must be far more of it than there is ordinary, visible matter. Physicists have made several suggestions about what could make up this dark matter, but so far no one knows for sure.

Dark energy: Pushing the universe apart

Since Hubble discovered the expansion of the universe, most scientists have believed that the cosmological constant was zero, which means that the expansion rate of the universe has steadied. But recent findings have indicated that the expansion rate of the universe is in fact increasing, meaning that the cosmological constant has a positive value. This repulsive gravity — or dark energy — is pushing the universe apart.

*   *   *

The above shows that galaxies are not behaving as they should and the universe is continuing to expand at an accelerated rate. Something appears to be holding too-quickly spinning galaxies together, something we cannot see or detect. And something else seems to be shoving them further away from each other at an ever-quickening rate.

But should this mean a search for more speculatory particles with dark matter? Or dark energy, some kind of reverse gravity, from the unknown processes of more undetected speculatory particles?

I would suggest, in accordance with the simplicity of the unified relativity that this book represents, that the first assumption should be that there *are* no other particles than protons; that our puzzlement is due to our lack of understanding of exactly what we're looking at across such expanses of spacetime – not our lack of identification of more hypothetical particles.

Conclusion

"Sometimes attaining the deepest familiarity with a question is our best substitute for actually having the answer." - Brian Greene, *The Elegant Universe*

In that sentence, I think that Brian Greene has summarized the attitude of many physicists, especially those whose training and experience has driven them into an over-familiarity with all the unanswered questions in quantum mechanics. The standard model proposed too many inexplicable phenomena, with which too many physicists have become too complacently over-familiar. Quantum mechanics and classical physics just cannot be reconciled, too many say. Which would be fine if quantum mechanics then went on to explain everything that classical physics could not, but it doesn't. It attains the deepest familiarity with the questions and presents that as an alternative to having any answers.

But Unified Absolute Relativity has answered all the outstanding questions. Unified Absolute Relativity shows that, with every atom a centre-point of curved space whose boundary only ever exists with respect to the next atom, nothing exists nor needs to exist beyond these fundamental particles to create the whole of our one and only, wonderful universe. The space between these particles *is* time passing, so if we accelerate one with respect to the other, it is small wonder that time passes at a different rate, according to Einstein's Special Relativity. But it is these accelerations which result in the field effects of light and the whole spectrum of electromagnetism. Atoms respond to accelerated (excited) atoms, with no "carrier" particles in between.

A very interesting and I think plausible theory is that as the universe continues to expand, eventually all particles will have dispersed, with the same distance between each. Although the universe may continue to expand, along with the distances between particles, each particle will still be separated from the next by a similar distance. As there will be no relative changes, nothing will happen. This has been called:

The Heat Death of the Universe

This idea suggests that as everything is evenly distributed, everything remains in stasis. But the particles all still sit there. But I would go further: there is a limit to the expansion of the universe and this will be reached when each particle is so far from every other, it cannot be gauged to exist.

Current thinking does not allow this idea, suggesting instead that each particle will still retain its ground state. That is the energy at which it sits, at rest. But that ground state must be measured from somewhere. In the here and now, in our world, that ground state is measured from the "ground", as it were; that is, from the reference point of everything else. But the everything else in that far-away expanded universe consists only of individual, massively separated single particles. In other words, the energy level, the very existence of things must be measured between things – and if the only things that exist are so far from one another their mutual effect is zero, then their ground state is zero. They have gone, in effect. The energy that is our universe will have dispelled.

But that state of far-away nothingness at the end of our universe simply mirrors the far-away nothingness at the beginning of our universe; as something to nothing will go, something from nothing may very well come, once again.

www.ingramcontent.com/pod-product-compliance
Lightning Source LLC
Chambersburg PA
CBHW032311210326
41520CB00047B/2883